The Place of the Humanities in Medicine

**INSTITUTE OF
SOCIETY, ETHICS AND
THE LIFE
SCIENCES**
THE HASTINGS CENTER

The Place of the Humanities in Medicine

ERIC J. CASSELL, M.D.

Copyright © 1984 by the Institute of Society, Ethics and the Life Sciences

All rights reserved.
No part of this book may be reproduced or transmitted in any form or by any means, electronic or mechanical, including photocopying, recording or by any information storage and retrieval system, without permission in writing from the Publisher.

The Hastings Center
Institute of Society, Ethics and the Life Sciences
360 Broadway
Hastings-on-Hudson, New York 10706

Library of Congress Cataloging in Publication Data

Cassell, Eric J., 1928-
    The Place of the humanities in medicine.

    Bibliography: p.
    1. Medicine—Philosophy.   2. Medicine and the humanities—Study and teaching.   3. Medical education.
I. Title.
R723.C4283   1984   610'.1   84-9139
ISBN 0-916558-19-3

Printed in the United States of America

Contents

Introduction	6
Earlier Relationships Between the Humanities and Medicine	8
The Present Situation	12
What the Humanities Have to Offer Medicine	15
Putting Living Persons Back into Medicine	16
Teaching the Humanities in Medical Schools	19
Literature	21
History	24
Understanding Time and Process	27
Philosophy	29
Teaching the Tools of Thinking	30
The Workings of the Word	33
Learning to Reason	40
Obstacles to a Role for the Humanities in Medicine Today	43
New Challenges for the Humanities	50
Conclusion	54
Bibliography	55

Eric J. Cassell

Eric J. Cassell is Clinical Professor of Public Health at Cornell University Medical College and Director of the Cornell Program for the Study of Ethics and Values in Medicine. He received his B.A. from Queens College, his M.A. in Biology from Columbia University, and his M.D. from New York University. A member of the Institute of Medicine of the National Academy of Science, he is a diplomate in internal medicine and a practicing physician. For more than a decade he has been studying and writing about the art of medicine and working on a theory of clinical medicine. He is the author of *The Healer's Art* (Lippincott, 1976; Penguin 1979), and a forthcoming two-volume work on doctor-patient communication, *Talking With Patients* (MIT Press, 1984). In addition to many early papers on the epidemiology of air pollution, he has contributed numerous articles to professional books and journals on medical ethics, the care of the dying, and the theory of clinical medicine. This monograph was written during the tenure of a Sustained Development Award from the National Science Foundation and the National Endowment for the Humanities, and a grant from The Commonwealth Fund.

Preface

This monograph by Eric J. Cassell, M.D., was prepared as part of a project carried out with the support of a grant from the National Endowment for the Humanities. The project of which it was a part was devoted to the question of whether and in what ways the humanities can meaningfully and effectively be addressed to urgent and immediate issues of public and social concern. Can there be, as we formulated the question, an "applied" humanities?

There can be little doubt that the field of biomedicine has exerted a special attraction to those in the humanities. The enormously increasing power of medicine to change individual lives, to improve life expectancies, and to profoundly influence social policy have all proved rich fare for philosophical, historical, and literary examination, interpretation, and analysis. A proliferation of books and articles, courses and lectures, comment and criticism, have been an obvious mark of the interest.

While those from the humanities drawn to biomedical issues, problems, and themes have not necessarily needed the prodding of physicians, biologists, and health care workers, they have nonetheless had it. As Dr. Cassell makes clear, many in the field of medicine would welcome a stronger role for the humanities; but he also makes clear that the response to that need has not been consistently positive. He offers a number of suggestions for a more fruitful relationship.

In addition to this monograph, the study of which it was a part also produced a general examination of the humanities *On the Uses of the Humanities: Vision and Application* (Hastings-on-Hudson, N.Y.: The Hastings Center, 1984). A forthcoming book, *Applying the Humanities*, edited by us, will be published in 1985 by Plenum Press and will present a number of papers on the humanities and public life, and the humanities and public policy.

DANIEL CALLAHAN   ARTHUR L. CAPLAN   BRUCE JENNINGS
Project Co-Directors

Introduction

This essay deals with the place of the humanities in medicine, past, present, and future. They are considered here to encompass not only the major disciplines of history, languages, literature, and philosophy, but also those disciplines not included in the natural or social sciences that express human values and the spirit of humankind. My general thesis is that the humanities have always had a place in medicine, and that they will play an increasingly important, necessary, and specific role as medicine evolves beyond its present romance with technology toward a more balanced view of the origin and treatment of illness.

To support this argument, I am going to discuss the classic view of the place of the humanities in medicine, describe existing programs in medical schools, examine what the humanities have to offer medicine, and then consider the obstacles to continued or increased participation by the humanities. Finally, I am going to discuss a shift that is occurring in medicine toward a primary concern for sick *persons*, instead of disease alone, and show how the humanities can support and advance this trend. Here, I shall argue that medicine has demands to make of the humanities that may exceed what humanists presently have the desire or knowledge to offer. Thus, in the process of meeting the changing needs of medicine, the humanities' view of their own nature and function may change.

This discussion is based on certain assumptions about medicine. The first is that medicine is about the care of the sick; everything else is secondary to this goal. The second is that doctors treat patients, not diseases, and as a corollary, that all medical care flows through the relationship between doctor and patient. My final assumption is that, for doctors, the body has primacy.

I believe the changes in medicine that are occurring today are part of a larger social upheaval that has been underway for more than twenty years. This social movement includes an increasing stress on what might be called personalized (even egocentric) individualism—more intense than the political individualism of our past—and is marked by a turning away from science and technology—even, on occasion, from reason itself (Olan, 1977, Doi, 1981). Among the effects on medicine of this rejection of Western scientific thinking have been an affinity for the "natural," and the embrace of a vast array of "alternate" therapies, from acupuncture to Zen, in place of scientific medicine. But with time it will become apparent again that science and technology are not the enemies; and there will be a more widespread understanding that "reason" is not *inherently* atomistic or reductionist, nor science the enemy of persons. Then the search for the solutions to the problems faced by medicine, as it turns toward new definitions of its role, will inevitably involve the development of new and exciting intellectual tools. I believe that in the coming decades the humanities will find themselves increasingly engaged in this leading edge of medical progress.

Earlier Relationships Between the Humanities and Medicine

Medicine is one of the oldest "learned" professions. From Maimonides to Lewis Thomas, the history of medicine is studded with the achievements of illustrious physicians who were broadly educated and cultured. As idealized over the ages, a physician is someone who knows more than merely recipes for the sick and has learned more than just the sciences of the body. At present, medicine is seen by both doctors and the public as indissolubly and exclusively linked to science, but this has not always been the case. In mid-nineteenth-century America it was believed that the physician's knowledge should include "literature of the past and present; history ancient and modern; politics, and religious beliefs of the times." (Dr. William Fussell in an address to the graduating class of the Woman's Medical College, 1857. Quoted by Ruth Abrams, mss., 1982). In his book, *Doctor and Patient*, S. Weir Mitchell, a famous American doctor at the turn of the century, similarly recommends that physicians expand their literary, artistic, and nonmedical scientific interests; he concludes: ". . . he will not be the worse, but the better, physician for such enlargement of his pursuits as I refer to, for we may feel sure that in my profession there is room for the direct or indirect use of every possible accomplishment."

In this view, a liberal education will increase physicians' understanding of and affection for the human condition. During this era, the latter half of the nineteenth century, the healing role of physicians was emphasized. Consonant with a sometimes almost ministerial view of the profession, in 1900 18 percent of doctors in Boston were women. Thus, the humanities were a natural ally of a medicine considering itself to *be* a humanity. As the century came to an end, however, the importance and force of science was increasingly felt, as science started toward its present overwhelming intellectual dominance of the profession.

A somewhat less charitable explanation of why, historically, physicians have been broadly educated comes from the sociology of professions. In this interpretation, an aristocratic model limits access to membership, keeping medicine's product scarce and valuable. In nineteenth-century Britain, those members who did not come from the elite moved into the higher orders of the profession by virtue of education. In 1847 a spokesman for the Royal College of Physicians commented on "the great advantages which result to society from there being an order of men within the profession who have had an education with the members of the other learned professions; from a certain class of the medical profession having been educated with the gentry of the country and having thereby acquired a tone of feeling which is very beneficial to the profession as a whole" (Larson, 1977). The education referred to was acquired at the great British universities; it involved classical learning, not "useful knowledge."

In the view of William Osler (who, probably more than any other physician, provided the model for the current form of medical education), to be a good physician *required* that one be broadly read and educated. Osler so embodied this ideal that the editors of his scholarly Silliman Lectureship at Yale, on "The Evolution of Modern Medicine," could state that they "have no hesitation in presenting these lectures to the profession and to the reading public as one of the most characteristic productions of the best-balanced, best-equipped, most sagacious and most lovable of all modern physicians" (Osler, 1919).

Unfortunately, the ideal represented by Osler, and the other views of medicine noted above, were rarely found in the United States during the latter half of the nineteenth century or the beginning of the twentieth. At that time, physicians did not constitute an educated elite; instead, American medicine consisted primarily of poorly trained doctors with competing medical ideologies, such as homeopathy, allopathy, and osteopathy. Educated men, and a few women, were trained at the small number of good medical colleges, but many more doctors were graduated from proprietary schools, night schools, and short-lived medical institutions with few educational requirements. Many physicians were trained entirely by preceptorships.

The greatest influence upon American medicine at the time, however, was probably German medicine, and here, education was extremely important. In contrast to more aristocratic British notions, however, this education was related to the concept of *bildung*—self-cultivation. In part as a result of the German influence, the primacy of the role of *education* in medical education was emphasized in the ensuing reformation of American medicine. Medical education was not seen merely as an initiation into a professional structure in conjunction with the acquisition of medical skills, but rather as a pursuit in which the intellect should be dominant.

Abraham Flexner's report to the Carnegie Foundation, recommending broad changes in the system of medical education, had a profound and widespread effect when it was issued. It is often assumed that the force of the Flexner report resulted from the disorganization and backwardness of American medicine at the turn of the century. However, I believe that the shifting emphasis of the medical ideal away from a reliance on the humanities and toward a fundamental alliance with science was also contributory (Flexner, 1910). This is reflected in that a basic tenet of the Flexner report was that medical care should be thoroughly grounded in the sciences. It followed that medical education should heavily stress biomedical sciences. For this to be practical, students needed some knowledge of the sciences on entering medical school. As a result of the Flexner Report, proprietary medical schools disappeared, many medical schools became affiliated with universities, and some college training was required for entering medical students.

Always a concern, early in the post-Flexner era the debate over the proper educational background for physicians resumed, and the tension between humanism and scientism has continued ever since. As Bruer has shown, commissions and reports, from the late 1920s to the present era, have championed liberal education over narrow vocationalism. Nonetheless, continuing surveys of medical school requirements testify to the lack of impact of such recommendations (Bruer, 1980).

George Engel has argued that Flexner meant to train medical students in the sciences in order to inculcate scientific thinking—forming hypotheses, gathering data, testing the hypotheses, and so on—rather than primarily to communicate scientific facts; Flexner's proposals, however, did not have this effect (Engel, 1978). When the interpretations (or misinterpretations) of the Flexner report were joined by a marked increase in the number of medical school applicants, the result was the all-too-familiar "premed syndrome": enormously competitive students fighting for admission to medical school, concerned with grades almost to the exclusion of any other educational goals, and concentrating almost entirely on the sciences in their premedical curriculum.

It is not surprising that such students focus on the factual content of courses necessary to perform successfully in exams. Unfortunately, they frequently do not acquire sufficient grounding in basic theory and scientific method. They also lack self-motivation for the continuing education that is a requirement of professional life in medicine. Because of their belief that medical schools want only applicants who have majored in science, such students avoid humanities courses, or take only required courses or those that can be expected to provide a good grade with little effort. Another factor that discourages premedical students from courses in the humanities is the necessity for studying sciences early in their college career in order to prepare for the Medical College Aptitude Tests (MCATS). Dr. Lewis Thomas has commented that:

> The influence of the modern medical school on liberal-arts education in this country over the last decade has been baleful and malign, nothing less. The admission policies of the medical schools are at the root of the trouble. If something is not done quickly to change those policies, all the joy of going to college will have been destroyed, not just for that growing majority of undergraduate students who draw breath only to become doctors, but for everyone else—all the students and the faculty members, as well.

In addition to the effects of a scientific bias on the part of a large number of medical school applicants, the difficulties for the humanities have been compounded by the national shift toward education in the sciences following the first Russian space flight, and the current shortage of funds for education in general.

> "The influence of the modern medical school on liberal arts education in this country over the last decade has been baleful and malign, nothing less."

The Present Situation

From the nadir of the influence of the humanities on medical education of just a few decades ago to the present time there have been major changes. At present, the majority of American medical schools offer some form of instruction in the humanities. This varies from the full-fledged departments found in only four schools (Pennsylvania State University College of Medicine, Wright State University School of Medicine, Southern Illinois University School of Medicine, and the University of Nebraska College of Medicine) to the innumerable "programs" that have their home in other departments or in deans' offices. There are, in addition, several schools that have institutes that are free-standing but closely attached to their schools of medicine.

The word "program" needs to be understood. In medical schools, any faculty or student effort regardless of size, whose existence is officially recognized, may be called a "program." Thus, a program may consist of an unfunded idea for teaching, research, or service which claims the attention of a faculty member and which has the dean's approval, even if nothing is being done as part of the program. Because designating something a "program" implies that the school recognizes its importance, it is assumed that "programs" will acquire funding, grow, and perform their function. For that reason, a "program" may also be well funded, fully staffed, and extremely productive. The Society for Health and Human Values periodically publishes reports on the state of human values teaching in the United States. Their most recent summary, *Human Values Teaching Programs for Health Professionals*, published in 1981, should be consulted for details (McElhinney, 1981). The Institute of Human Values in Medicine and its staff maintain an active interest in this area and constitute the single most valuable information resource.

With the exception of departments of the history of medicine, some of which have been in existence for many years, these programs began to come into being in the 1960s. Why this development occurred is not entirely clear, but it was, I believe, initiated primarily by students. Many who taught during that era may still get hives when they hear the word "relevance." Students were no longer content to be taught what their faculty believed important. It was essential to the students that their classes be "relevant" to the problems of poverty, racial bias, and political "oppression." These issues were perceived by many students to be widespread in the society with which they were frequently disillusioned. Medical school faculty, eager to avoid the strife that at that time embroiled colleges and universities, were quick to invite students to participate in the planning of the curriculum and the choice of "relevant" studies. Further, informal classes were often planned and taught by students (and invited faculty) in subjects that stressed the humane aspects of medicine and medicine's commitment to the poor, the neglected, and the downtrodden. What came out of this was the breakdown in the rigid hierarchy of relations between students and faculty, along with the voice of students in the curriculum. The concern that medicine be more humanistic (a word employed more often in this era) also persisted, as did some of the courses that were begun at that time to teach the humanities. But because students were frequently more clear-sighted about the overly scientific, depersonalized, and hierarchical medicine to which they were opposed than they were clear-thinking about what was to take its place, most of the innovative educational ventures of that period did not survive. The revolt against "science-and-technology," which had wide support among students in the sixties, did not have a set of alernate concepts to put in place *that would be effective in the care of the sick.* Such effective alternatives are just beginning to come along, aided by the currently widespread teaching of humanities in medical schools. A note of caution is necessary because, although enthusiasm for the humanities continued into the 1970s, there has been some slowing of growth recently. This may reflect a fall in the amount of money available from grants and the institutions themselves, but it might also represent what Daniel Callahan has termed the "ethics backlash." Although there may be elements that in combination have diminished the continued rapid growth of administratively strong programs, there is no sign of waning interest on the part of involved faculty or the students. Let me illustrate with a description of a few programs:

Pennsylvania State University—Hershey Medical Center. Program founded in 1967 (with the start of the medical center). This program has full departmental status, five full-time faculty supported by the university budget, and is completely integrated into the

> . . . although enthusiasm for the humanities continued into the 1970s, there has been some slowing of growth recently.

medical center functional and governance structure. Students are required to take at least two courses in the department during their medical school career. Courses are offered in death and dying, medicine and ethics, major medical novels, religion and medicine, philosophy of medicine, history and philosophy of genetics, and many more.

State University of New York at Buffalo. Program founded in 1977. The Committee on Human Values and Medical Ethics is located in the dean's office from which it receives its funding. There are no full-time faculty. Teachers are drawn from those in the medical school who are committed to the program, as well as supporters in other schools and departments in the university. The committee offers an elective course for first-year medical students in addition to various lecture programs offered university-wide.

Yale University. No formal program in the medical school and no medical school faculty. Courses in such things as medical ethics, medicine and the law, and similar topics are given by other appropriate schools of the university.

Cornell University Medical College. Founded in 1979 as the Program for the Study of Ethics and Values in Medicine in the Department of Public Health. One faculty member supported primarily by grant funds. Elective course given in each of the four years.

As reported by the Institute of Human Values, four schools have departments of humanities, seven have departments of the history of medicine, three schools have divisions within other departments, twenty-three have programs located within other departments, eighteen medical schools have their humanities programs located in the dean's office, and fourteen others have a variety of administrative structures. It is important to be aware that many programs included under the rubric "humanities" are primarily, if not exclusively, concerned with teaching ethics.

What the Humanities Have to Offer Medicine

Putting Living Persons Back into Medicine

> ... physicians are trained from the first days of medical school to disregard the knowledge they bring with them of everyday life and human function as irrelevant to medicine.

Ideally, physicians should define their diagnostic and therapeutic goals in terms of the everyday life and function of individual patients. Unfortunately, that ideal is seldom met because of the difficulty of holding impersonal technical imperatives in check, and because doctors seem to be trained to focus on diseases almost to the exclusion of how sick persons actually live their lives in families and communities. In part the problem arises because physicians are trained from the first days of medical school to disregard the knowledge they bring with them of everyday life and human function as irrelevant to medicine. Another obstacle is that doctors are not trained to include in their decision making the kind of "soft" and often subjective information that is relevant to the everyday life and function of sick persons. Correction of these educational errors would do much to help change physicians' priorities in patient care.

I believe that education that teaches students to apply knowledge of disease and the body to persons in their everyday life and function takes advantage of students' preexisting knowledge of the world, as well as what they have learned of medical science. Teaching them how to acquire information about persons expands the students' knowledge and broadens the basis for the application of basic science and pathophysiology. Teaching physicians how to acquire, evaluate, measure, synthesize, and analyze information about sick persons and their bodies provides them with the tools to gain the information about individual patients necessary to meet the ideal of care. Teaching doctors how to *think* about both the body and about persons—about objective and subjective, data and values, analytically and synthetically—allows them, in formulating the goals of patient care, to integrate their knowledge of medical science, the body, and the everyday life and function of individual sick persons. In recent years, the Association of American Medical Colleges has endorsed the use of the teaching of humanities to broaden students' perspectives with an eye to these ideals.

While meeting these goals is desirable, it should be noted that the distance that physicians' training places between them and sick persons is a *necessary* component of medical education. Another word for "distancing" the patient is depersonalization. In order to teach medical students the sciences of medicine, it is necessary to depersonalize the human body for them. Their cadavers, experimental physiology, pathology, indeed, all the preclinical studies provide a dehumanized arena in which to learn human biology. This aspect of medical education is, I believe, an *essential* part of the socialization of physicians. Students *must* learn to conceive of the body as a thing apart from and different from their own bodies. (John Cody, who took the course in gross anatomy twice, first as a medical artist and then as a medical student, provides an interesting description of how the student dehumanizes the cadaver in the beginning of gross anatomy. Cody, 1978.) In this regard the students' experience recapitulates the history of medicine. The anatomical drawings of the seventeenth century, no matter what they are meant to demonstrate, show the dissected body in a personalized position—reclining comfortably, an arm uplifted as if pointing, or in some similar pose. This is in sharp contrast to anatomical demonstrations of the last hundred years, in which no trace of personalization is to be found. For example, a series of seventeenth-century anatomical drawings, meant to explore the leg as a mechanical device, not only portrays the legs connected to a personified body, but to demonstrate that the leg is weight bearing, shows the figure carrying a globe of the earth, Atlaslike. A series of 1982 studies of the mechanics of the legs contains drawings in which the representations of the legs bear no resemblance to actual legs and are not connected to anything bodylike. The body has yielded its secrets in a consistent manner only since experimental and statistical methods were developed that *totally* divorce scientific generalizations from the individuality of persons. Well-trained physicians are able to think in body terms—"think" heart, "think" kidney—completely apart from the fact that the kidney in question is John Smith's kidney, and that what they are thinking is bad news for John Smith. On the other hand, effective physicians are able to bring the person, John Smith, back into the picture and make decisions that integrate facts about both the depersonalized kidney and the individual patient.

Superb training in the sciences of medicine is the hallmark of modern medical education. Unfortunately, neither the need for *repersonalizing* the physician's knowledge nor methods for teaching the ability to apply the generalities of science to individual patients have made much headway in medical schools. There is

The body has yielded its secrets in a consistent manner only since experimental and statistical methods were developed that *totally* divorce scientific generalizations from the individuality of persons.

reason to believe that training in the humanities might lead to these goals.

But what must be taught? That doctors take care of sick persons, not just their diseases? Why is not the saying of it the equivalent of teaching it? Because the problem to be overcome is not simply that the education of the medical students has dehumanized their knowledge, but that in every encounter with a patient, the knowledge of the body that guides the examination—whether it be history taking or physical examination—and the categories used by physicians to process the forthcoming information, push the person of the patient into the background. Further, and equally important to remember, as physicians are working they have no awareness that their actions and their modes of thought continually brush aside the person. Indeed, especially in the present era, they usually have no wish to do this, nor do they think of themselves as acting in such a manner. Since in common with everyone else, they have not been taught how to step out of themselves and watch their own actions, they cannot appreciate the impact of their words and manner. (When medical students or physicians see videotapes of themselves with patients, they are often startled and even dismayed at how unaware they were of their behavior and its impact.)

The problem that is created when a person is dealt with as irrelevant in the care of his or her own sick body is related to and amplified by the fact that the American culture has largely excluded the body from everyday awareness, except perhaps for recognition of its appetites and its liability to damage. This observation remains true despite the recent upsurge in interest in sports.

> . . . as physicians are working they have no awareness that their actions and their modes of thought continually brush aside the person.

Teaching the Humanities in Medical Schools

The first question that must be faced by all who teach the humanities in medical schools is whether they are there to teach medical students or to teach their individual subjects. I believe the needs of the medical student require that the teaching of the humanities be shaped to special intellectual tasks. It is a necessary area of inquiry to find out exactly what is unique about these needs. It is my hope that this monograph will suggest some areas of exploration. Scholars in the humanities who teach in medical schools are often dismayed when they discover how single-minded medical students can be. Students frequently have little patience for the intrinsic merit of humanities presented as part of the medical school curriculum. This disinterest does not speak to their lack of understanding, shallowness of interests, or the failure of their undergraduate medical education. For many years, medical students have been the brightest, most capable, most strongly motivated, hardest-working young people that one could hope to teach. They are aware of the enormity of the responsibility that will be theirs all too soon. This generates an impatience with the humanities, when students do not perceive them as adding to the store of knowledge required to be a good doctor—what their faculty believes to be of central importance, as judged by examination material, departmental curriculum time and relative status of faculty. They are interested in *medicine*, and the burden of proof that something is relevant rests with the instructor. Occasionally, a teacher will rebel against their medical monomania, insisting that, say, they come to the Greek tragedies rather than that the Greek tragedies be brought to them. The students will usually stop coming to the class—as simple as that. K. Danner Clouser, speaking about teaching philosophy in a medical school, has said it very well:

> My basic awareness is that my role is to produce doctors and not philosophers. I don't even want to make them *mini*philosophers. Rather my goal is to make them philo-

The first question that must be faced by all who teach the humanities in medical schools is whether they are there to teach medical students or to teach their individual subjects.

sophical doctors. . . . I want to help students with medicine. . . . Using the skills and insights of these extramedical disciplines to *enhance* medicine and be *integrated with* medicine *is* the point (emphasis in the original) (Clouser, 1978).

Because medical students *are* so bright and motivated, when they are interested, they are a sheer pleasure to teach. However, one must understand that the word "doctor" is the name of a genus with many different species. Surgeons are different from physicians, and orthopedists are different from plastic surgeons. Pediatricians and neurologists can frequently be distinguished on the basis of their interests and, some believe, their personalities. Students intensely interested in these specialities tend to acquire the biases, interests, and colorations of their elders. These statements sound like biased stereotyping, but they have enough basis in fact to be of interest to nonphysicians in the medical school environment. Do not be dismayed, and do not take it personally, if a student who is headed for a career in neurosurgery or orthopedics finds little of interest in the humanities. The active disinterests of students with strong career goals are part of their socialization process and are best dealt with by employing the same humoring attitude that the other students display. Years later, however, it may be one of those same orthopedists who comes to tell you how important and memorable your course was. The basic idea is that just as you would want to know something of the indigenous culture if you taught in a strange land, so will teaching medical students be more successful, rewarding, and enjoyable when you learn about the culture of medicine. If you want to change the students' attitudes, devote your energy to changing their faculties' attitudes. Teachers in the humanities are frequently surprised to discover how much interest and enthusiasm a professor of, say, surgery has for teaching literature to medical students. Unfortunately, the enthusiasm of the faculty may fade during a hard-fought political battle in the curriculum committee, which is why wisdom suggests a posture of patient persistence in the pursuit of change.

> . . . just as you would want to know something of the indigenous culture if you taught in a strange land, so will teaching medical students be more successful . . . when you learn about the culture of medicine.

Literature

Literature offers the opportunity to see the interplay of illness and persons, the role of physicians in the lives of others, the impact of their own medical knowledge on the doctors' personal lives, and the perception of physicians by laypersons. Because literature is free of the constraints of the day-to-day world, it is able to offer a fuller picture of individuals, their relation to objects, events, and other persons, of the world of the sick and the meaning of illness to individuals, of how compassion, empathy, mercy, and other moral qualities are expressed and how they affect others. Joanne Trautmann has said it well:

> No matter how well trained as observers . . . we normally do not see the full range of details in any situation. What's more, as we go about our daily tasks, we do not have the time or the facility to arrange those details in some kind of memorably insightful patterns. Often we do not truly perceive even the familiar; and thus our knowledge of reality grows blunted and hazy. We move through unseen worlds.
> In contrast, the first-rate fictional world is a fully considered one. In it, lights are cast upon shadowy corners, or veils are stripped from dailiness. . . . Fictions are not bound, as medical studies are, by the actual patients who present themselves, the occasionally unreliable laboratory data, the regulations about human experimentation; . . . they can see the diseased person simultaneously from the outside of the body, the inside of the mind, and the experience of the doctor watching the diseased (Trautman, 1982).

The bibliography prepared by Joanne Trautman and Carol Pollard, *Literature and Medicine*, is a rich resource for those who wish to use literature in the teaching of medical students and need to isolate certain themes, or find materials that are most apposite to specific areas (Trautman and Pollard, 1975).

Kathryn Hunter has demonstrated, in her program at Morehouse, how the humanities (particularly literature) can reinforce the commitment of medical students to the goals of primary care (Hunter, 1982).

Sandra Bertman has used literature in a different sense, to affirm the legitimacy of students' own feelings in the painful interactions of medicine with human reality, such as death and loss, and to demonstrate that what concerns the physician has been important to humankind through the ages. In the arena of anatomy, she uses literature and the visual arts to explore the experience of dissection (Bertman, 1979, 1982; Marks and Bertram, 1980).

An interesting collection of essays, *Medicine and Literature*, edited by Enid Rhodes Peschel, gives an idea of the widely different ways in which literature can be used to illuminate the medical experience, and conversely, the manner in which medical experience has been used by writers who were physicians and those who were not. Anthony R. Moore, in *The Missing Medical Text*, demonstrates specifically how literature can be used to address aspects of physicianship not covered in the formal curriculum, to draw on the medical heritage that is recorded in literature, to heighten sensibilities for patient care and professional self-assessment, and to explore nonscientific ways of thinking and their importance in considering human issues. The book consists of short literary passages that illustrate specific topics—the patient's or the relatives' experience, portraits of doctors, ethical issues, and others—and a transcription of students' discussion following a reading of the materials. Not only are the discussions interesting in themselves, but they give someone who is unfamiliar with teaching in a medical milieu an idea of medical students' interests and concerns.

One area of teaching in which I do not believe literature has been used to best advantage is placing the body in the context of everyday life and function. As I noted earlier, except where sex or disease and death are concerned, our culture is relatively blind to the body—which seems largely to be conceived of as merely a vehicle for carrying around the head. To make the point, I remember my amazement on reading the Platonic dialogues and discovering the large part that the body and sexuality play in them. That aspect of Plato does not seem to enter the modern consciousness, where the body is considered an intrusion on the intellect. It might seem unnecessary to teach medical students about the body; after all what else is medicine about? In fact, however, while doctors know an enormous amount about how individual organs or body parts function and malfunction, they receive little training in how the body functions in everyday life. What practicing physicians do know, they learn from helping patients live their lives despite their disabilities. Elaine Scarry has discussed the place of the body and

> ... while doctors know an enormous amount about how individual organs or body parts function and malfunction, they receive little training in how the body functions in everyday life.

of work in the novels of Thomas Hardy to make clear both the centrality of the body in lived existence, and its virtual invisibility in the work of so many others. She points out that human consciousness, for Hardy, is always *embodied* consciousness, but for many other authors (and perhaps most of us) that is not the case (Scarry, 1983). Since the body cannot *not* be there, if it is not part of consciousness, it is being dismissed. It is difficult to conceive of a medicine primarily concerned with the care of sick (or well) persons in which the body, as a whole functioning entity, does not come to occupy a more central position. Here again, literature—but also the graphic arts—can best make the point.

History

Medical students, and often their faculty, tend to believe that the history of medicine began the day they were born. Consequently, their studies, their understanding of medical science, and their conception of the work of physicians have no historical context. This leads to difficulties that directly affect the function of clinicians and researchers. One problem arises because the depersonalization, which I noted was a necessary part of medical education, is exaggerated by the notions of medical science as timeless, objective, and value-free that have marked these last several decades (and which are now beginning to change). In a world view that excludes the importance of persons, history is not welcome. Conversely, one cannot highly value personhood without also acknowledging the fact that persons *always* have histories. Whole human beings can never be discussed solely in the present tense. Further, persons belong to families, and the family is an institution whose present is always indebted to the past and in thrall to the future.

As I pointed out earlier, it is only recently that medicine has begun to change toward a concern with the sick person as opposed merely to treating disease. It is important to point out that it is not only the person of the patient that has been absent from medicine during its recent history, but the person of the *physician* as well. Both patients and their physicians have come to act as though they believed that technology is what makes the diagnosis and cures the patient. In this belief, all physicians, to the extent that they know their medical science, are equivalent—individuals do not make a difference. This issue is made more complex because of a lack of synchronization in attitudes about the individuality of physicians versus persons in the mainstream of American society. From the 1930s through the 1940s, there was much comment, e.g., Charlie Chaplin's movie *Modern Times*, about the anomie imposed by the modern, increasingly technological, assembly-line industrial world. Later, the same kind of blurred individuality was denoted

> **Both patients and their physicians have come to act as though they believed that technology is what makes the diagnosis and cures the patient.**

by the image of the "corporation man," or the Madison Avenue man in the gray-flannel suit. During the same period, doctors were considered by many and believed themselves to be highly individualistic—whether they were clinicians or members of the newly emerging elite of research scientists, as described in Sinclair Lewis's novel, *Arrowsmith*. This is the individualism of effort that Americans classically associate with the western pioneers.

Since the 1950s, a new wave of individualism of a different kind has emerged in American society, characterized by its uniquely inner-directed and personal nature. This has been caricatured as a culture of narcissism by Christopher Lasch, in his book of the same name. But aside from the fact that it is an error to extend to a whole society an idea that was developed for the psyche of individuals, this change in self-concept should not be lightly dismissed. It constitutes, I believe, a major social change that will have enduring effects. Now, however, the dominant society, having returned to its more fundamental kind of individualism, and having turned away from the science and technology that saturates modern life, does not see *physicians* as distinct from *their* technology. This has led to the curious phenomenon of patients demanding that they be treated as persons at the same time that they depersonalize their physicians! But medicine and physicians are now also beginning to evolve to a point where doctors will begin to understand that, *qua* doctors, they are distinct from their tools and their science. It is toward that goal that projects such as teaching the humanities to medical students are directed.

. . . doctors will begin to understand that, qua doctors, they are distinct from their tools and their science.

Chester Burns has pointed out that during the forties and fifties, in contrast to previous and subsequent times, many medical schools offered courses in the history of medicine. He suggests a relationship between the popularity of the history of medicine and the pride of self and profession to be found during that era (Burns, 1978). The studies of Genevieve Miller on programs in the history of medicine in medical schools of the United States would seem to bear him out, although more complex forces are undoubtedly at play (Miller, 1969).

Another difficulty that stems from teaching medicine without historical referents is a kind of cohort egocentricity—a belief that what we do today is the best that has ever been done, that we have the most complete understanding of the human condition that has ever existed, and that (paradoxically) what we lack is just around the corner. In this view, history is merely the record of a never faltering climb to the (soon to be reached) summit of medical knowledge. When historical events are considered, they are judged by their contribution to the present. As Risse notes, this approach (called presentism by historians) effectively closes out any consideration of medicine before the mid-1800s (Risse, 1975). Another problem caused by presentism is that physicians fail to understand that in every era the profession and the work of individual physicians have been subject to forces generated by the interaction between medicine and society. Today, economic pressures and changing societal norms exert great pressure on medicine, and it would be helpful for physicians to realize that this has always been the case. Moreover, when the profession is fragmented into specialities and subspecialities, medical history provides a means of uniting it by emphasizing a common heritage (Shortt, 1982). According to Pedro Lain Entralgo, a methodical and carefully considered understanding of the history of medicine is fundamental to the development of newer philosophies and methodologies. He underscores the quotation attributed to the famous surgeon, Bilroth, who said that "History and research are so inseparably connected that one is for me unthinkable without the other" (Lain Entralgo, 1983).

Historians have agreed more about the importance of teaching history in medical school than about who should teach it and how. In the past, medical history was often the "hobby" of physicians with no special training as historians. This created tensions between the "amateur" historians and those with professional training in history. The resolution of these conflicts seemed at times to be more important to the contestants than medical history itself. Arguments over whether a medical background is necessary, or if the degree of academic specialization in history is needed are not easily resolved if one is trying to start an institute devoted to the history of medicine. But when it comes to instructing medical students, Temkin (quoted by Hudson) simply recommends a "good teacher who knows the subject." In this debate, Hudson emphasizes the importance of the quality of the teaching of history to medical students. He is aware that history is not seen as their most vital subject, and that their allegiance and attention must be *won* (Hudson, 1975).

> In the past, medical history was often the "hobby" of physicians with no special training as historians.

Understanding Time and Process

It never seems to occur to medical historians that anything but *medical* history should be part of the curriculum. While I believe that students benefit greatly from having a historical context in which to view their work and their science, history also has a methodology and an understanding of *time* and *process* that are unique among the humanities. This is true of all branches of history, and these two concepts play a vital role in clinical medicine. Central to taking care of the sick is understanding the history of their illnesses. Illnesses unfold over time, and in common with other histories they have contexts, a cast of characters (including, but rarely limited to, both the person and the body), occur at particular times and in particular places. Understanding the history of illness is a difficult task—even learning to present it well to other physicians takes a number of years. The problems arise because patients' stories of their illnesses are not dispassionate expositions of a set of "facts." Instead, events in the body that are considered symptoms (in itself, a relative category) are perceived, organized, and described according to patients' perceptions of their meanings. It is obvious that patients' and doctors' interpretations of those symptoms may be at variance—what is important to the physician may not be so to the patient and vice versa. Even conceptions of the passage of time are influenced by many factors that may interfere with the diagnostic utility of patients' recollections. Physicians taking the history from a patient must form hypotheses about the possible illness represented, but must be careful not to allow their hypothesis to influence the data. Further, doctors must learn to test their diagnostic hypotheses as they obtain further historical information. Despite the difficulties inherent in the process of taking the history of illness from an individual patient, history taking is rarely taught in any systematic manner in medical schools (although, there has lately been some evidence of increased interest). This lack is the more significant because, as has been repeatedly pointed out, 80 percent or more of diagnoses are made on the basis of the patient's history. Unfortunately, most physicians are inexpert at history taking—in years past, the most common reason that candidates failed their speciality board examinations in internal medicine was failure to take an adequate history! These facts suggest that historians might have something special to contribute to teaching how to obtain, understand, and use the illness history. I am aware that historians may not see this as either their *metier* or within their interests. Because illness is a process that takes place over time and presents itself in discrete, relatively short episodes, however, it lends itself to research on the dimension of time and the problems of thinking about process rather than events. When disease concepts as we know them evolved in the nineteenth century, diseases were considered to be things that invaded patients. The progressive explication of their structural basis, and more recently their biochemical basis, did not change the sense that diseases—whether cancer of the breast, tuberculosis, or even the "flu"—were "things" that produced an "event." The great medical advances of this century, however, have focused on pathophysiology: how the body works in disease as contrasted to its normal function. Despite its evident superiority, pathophysiological thinking has failed to produce its potential impact on day-to-day clinical action, in part because of the difficulty of thinking about processes unfolding over time as opposed to static states. Unfortunately, it is not clear how to teach physicians to think in process terms, or to deal with the dimension of real time. Because of the relevance of these issues to medicine, it seems to me that historians not only have something to learn from medicine, but something to contribute to it that is unique to their discipline.

Philosophy

Because of the growth of interest in ethics, philosophy is the humanist discipline that has made the greatest inroads in the medical curriculum during the present era. After more than two generations of dormancy, during which Otto Guttentag and Pedro Lain Entralgo represented almost the only voices directed consistently at the theoretical issues of medicine, the philosophy of medicine has become an active speciality with several journals and increasing numbers of publications. While ethical issues have received the greatest attention, philosophical problems related to the nature of disease and health; the fundamental bases of medical action; the relationship between doctor and patient; the nature of medical information and the problems of decision making and problem solving; the place of medicine in society; and many others have come under consideration in recent years. In their book, *A Philosophical Basis of Medical Practice*, Edmund Pellegrino and David Thomasma lay out many of the issues with which the philosophy of medicine must concern itself. For further information about the field, the reader should consult that text and the journals. Less attention has been paid to *clinical* theory—what Otto Guttentag might call the concerns of the attending physician—than to other aspects of the philosophy of medicine, but it is my hope that the theory of clinical medicine will receive increasing attention from physicians as well as philosophers. As Stephen Toulmin has made clear, philosophy not only has been important to medicine, but medicine has been a boost to philosophy in providing problems central to the human condition whose solution has been made urgent by the march of medical technology and the changing demands of society (Toulmin, 1982). Perhaps the same thing will be able to be said about other areas of the humanities within a few years.

The greatest philosophical emphasis in medicine has been on bioethics, and courses on medical ethics are now given in most medical schools in the United States. In many instances the enthusiasm of students has been responsible for initiating ethics in the curriculum, and their interest continues—often more strongly than that of their faculty.

Teaching the Tools of Thinking

Dr. Edmund Pellegrino has stressed the important distinction between the liberal arts and the humanities:

> It is a common misconception that the humanities and the liberal arts are synonymous. This is an error as common to humanists as it is to the general public. The liberal arts are attitudes of mind, not disciplines or bodies of knowledge. They have since classical times been those intellectual skills needed to be a free man—not only in the political sense, but more critically in the sense of being free of the tyranny of other men's thinking and opinions, free to make up one's own mind and take one's own position. The liberal arts comprise those skills most commonly associated with being human—the capability to think clearly and critically, to read and understand language, to write and speak clearly, to make moral judgements, to recognize the beautiful, and to possess a sense of the continuity between man's present and inherited past (Pellegrino, 1982).

Pellegrino has had more influence and written more, and more persuasively, on the subject of the humanities in medicine than perhaps any other contemporary thinker (Pellegrino, 1974, 1979, 1981).

William May has also discussed what the liberal arts have to offer. He takes the central tasks of the liberal arts to be understanding, interpretation, and criticism (May, 1982). Understanding is natural to modern medicine, where it is a matter of dogma that knowing what to do for the sick can only follow understanding the biological mechanism involved in the illness. Unfortunately, there has not been an equivalent stress on the need to understand the physician's act, the relationship between doctor and patient, or other nonscience areas of medicine. Interpretation and criticism have yet to have their full force. Interpretation involves examining what is or has been "significant, humanly important, and worthy of report." This activity differs from the scientist's reexamination of previous theories or research precisely because interpretation involves the introduction of value perceptions—was a particular medical concern or involvement worthy of the effort, and if not, what should or could have been changed? The inevitable "restless revision" and changing interpretation of such scrutiny would be unsettling to physicians. Interpretation not only keeps the past alive, but keeps the present under a light of interpretative examination generally absent from medicine—except when something has gone wrong. Criticism would also seem to be an inherent aspect of the best of medical science. Surely scientists are always poking and probing in order to refine the methodologies and hypotheses on which scientific progress depends. Less often is there an *established* form by which criticism of the whole medical endeavor, or its parts, is deemed an important and essential aspect of medicine itself—a respectable pursuit of physicians trained in critical thought—rather than mere scholastic carping or lofty words of wisdom offered by an aged medical savant before leaving the profession. The work of critics, arising from within and respected by the medical profession, might well keep medicine more finely tuned to the surrounding society. Kathryn Hunter has suggested both the feasibility and utility of looking at medicine with eyes trained in the study of literary texts. The behavior of physicians, questions of courage and heroism, the relationship of therapeutic distance to aesthetic distance are all subjects that might benefit from the tools of literary criticism. She has pointed out that medicine, which officially shuns the anecdotal, is awash in stories—stories used as teaching devices, as the social glue between staff, and as cautionary tales, to name just a few of their functions—all ripe for examination as texts.

However, the liberal arts have something special to offer physicians, if sick persons, rather than their diseases, are to be the new focus of medicine. Fresh concepts, skills and guidelines for behavior are necessary to supplement the strictly technical. In medicine, we do not simply describe the procedure for an appendectomy and then leave students to their own devices. We define appendicitis, base the definition on anatomy and pathology, demonstrate how it manifests itself, how the diagnosis is made, and how to treat it. Similarly, it is not sufficient to tell a student or physician to treat sick persons, not just their diseases. Without the necessary definitions, tools, and skills, all that has been created is a moral injunction: "Go and do thou likewise." When a person fails to fulfill a moral injunction, he or she generally ends up taking the blame and feeling badly. Then, despite good intentions, patients are not better

> . . . the liberal arts have something special to offer physicians, if sick persons, rather than their diseases, are to be the new focus of medicine.

off and doctors feel like failures. This is frequently what happens nowadays when physicians begin their internships. They are supposed to care about their patients as "persons," use their "feelings," and be "open" and communicative. Given the facts that they have no specific training in this aspect of patient care, that they are overwhelmed by work, and that they are usually rewarded for technological rather than interpersonal skills, their sense of inadequacy may defeat their good intentions.

From time immemorial there have been physicians who were extraordinarily adept at working with patients—taking histories, establishing rapport, achieving compliance with even the most unpleasant regimens, being sensitive to unspoken needs, providing empathetic support, communicating effectively, and even getting paid after the illness. This expertise, usually called "the art of medicine," is generally acquired after years of experience. Some doctors, nonetheless, are more skilled with patients than others. Because of this, it is frequently said that the art of medicine is tacit knowledge, a matter of "intuition," and that it is unteachable: "You either have it or you don't." (This last heard, strangely, as often from humanists as scientists.) I am convinced that the art of medicine can be taught and studied in a systematic and disciplined manner. Critics from within medicine often act as though the words "systematic and disciplined" are applicable only to science and are incompatible with "art." Humanists would be quick to dispute the issue pointing out that, say, Ludwig van Beethoven, judging from his notebooks, was extremely systematic and disciplined, as was Michelangelo, Pablo Picasso, and probably almost every good artist. Regarding the possibility of teaching an art, while it may be the case that talent or intuition is a feature, even child prodigies have teachers and work constantly to refine their skills. In the absence of disciplined effort, they would surely not realize their promise. In an essay such as this, it is sad to have to make a case for the fact that an art can be taught.

The art of medicine is composed of abilities in four different but interrelated areas. The first is the ability to acquire and integrate subjective and objective information to make decisions in the best interests of the patient. The second is the ability to strengthen and utilize the relationship between doctor and patient for therapeutic ends. The third is knowing how sick persons (and doctors) behave. Finally, the central skill upon which all the others depend is effective communication.

> I am convinced that the art of medicine can be taught and studied in a systematic and disciplined manner.

The Workings of the Word

In the practice of medicine, communication skills are essential to the diagnostic and therapeutic process. This is true in the social sense, that is, doctors should be able to establish their interest and concern for their patients and be attentive to patients' problems and anxieties. Indeed, the most common complaint that patients have about doctors is that they do not listen.

There is a deeper sense, however, in which doctors need to know how to speak and listen effectively. Speakers literally portray themselves when they speak. To tell someone about objects, events, or relationships is to tell the attentive listener about yourself. The choice of adjectives, adverbs, verbs, nouns, and pronouns "places" the speaker in relation to what is being described. To say, "I went to Doctor Jones with a pain in the chest and he stuck me right in the hospital," is quite different from, "I went to Doctor Jones when I had pain in my chest and he thought I should go right to the hospital." Although the events described are the same, the first speaker has distanced himself from the pain and described himself as more passive to the physician's action than has the second. If we believe that the nature of the person modifies the illness in its presentation, course, treatment, and outcome, then the kind of information that a patient's speech offers can be of importance to the physician. The spoken language is the most important tool in medicine; almost no diagnostic or therapeutic act occurs in its absence.

Given its importance in medicine, one would think that spoken language would have been subject to intense scrutiny. Yet the systematic study of natural conversation, in or out of the medical setting, is a relatively new discipline. For science, the basic difficulty derives from the fact that language differs fundamentally and irreconcilably from other objects of scientific inquiry. Science has been successful because of the ability of scientists to study, in controlled isolation, simple, linear, cause-and-effect parts of more complex wholes. This produces "dyadic" statements of the type

> **The spoken language is the most important tool in medicine; almost no diagnostic or therapeutic act occurs in its absence.**

with which we are all familiar: "If A, then B." Language, however, is totally different: it is irreducibly triadic. Words do not merely stand for specific objects, as in "Apple is a word that stands for the firm, fleshy, edible fruit of the tree, *pyrus malus*." How about, "She's the apple of my eye," or, "One bad apple spoils the whole barrel," or even, "Adam's apple." Words always stand for a particular thing *to someone*. The irreducible triangle consists of a word, the thing it stands for, and the person for whom it has that meaning (Percy, 1975). Complications arise because words can mean different things to different people. Since one cannot verify in "objective" terms what is going on inside another person's mind, the problem of personal meaning has proved impenetrable to "hard science." This does not trouble humanists in the slightest—indeed, the scientist's dilemma provides a measure of disciplinary comfort to humanists who may bridle at the thought of an "objective" or "scientific" approach to the problem of meaning.

> **Each illness caused by a given disease is unique, however, and differs from every other illness episode because of the person in whom it occurs.**

Fortunately, all the features that make the spoken language opaque to science provide wonderful opportunities for clinicians. Human illness, in fact, is triadic in the same manner as language. Diseases, when isolated and confined to their afflicted cells, organs, or enzyme systems, are quite constant in the manner in which they express themselves, and we have instruments that measure their activity, just as a thermometer measures the kinetic behavior of molecules. Each *illness* caused by a given disease is unique, however, and differs from every other illness episode because of the *person* in whom it occurs. Even when a disease recurs in the same person, the illness is changed by the fact that it *is* a recurrence; it now carries the associations and the history of the previous episode. (This effect of repetition—that something is the same, but now both changed and enlarged by virtue having been there before—is used to effect in both music and literature, especially poetry but also fiction. The employment of repetition in the arts mirrors its occurrence in the world.) While it is obvious that genetic makeup or changes in immune response can alter the reaction to disease, as can diet, personal habits, and level of physical conditioning, the presentation, course, and outcome of a disease can also be affected by whether the patient likes or fears physicians, "believes" in medication or abuses drugs, is brave or cowardly, is "self-destructive" or vain, has unconscious conflicts into which the illness does or does not fit, and so on. These features are part of the illness, for illness is not only a physical event, but a "meaning event" as well. Indeed, there is no event that befalls humans to which meaning is not attached. It is the triadic nature of human illness that makes the art of medicine so vital; if every patient were the same, then merely to know the disease would be to know the illness.

Despite the obvious importance of spoken communication, there are few medical schools that teach more than the rudiments of taking a history. None, to my knowledge, have understood the basic importance of spoken language and attempted to teach not only the effective *use* of medical rhetoric, but its *fundamentals*, in the same way that the fundamentals that underlie, say, methods of diagnosis are taught. Even among those in the humanities, sadly, skills in using spoken language have fallen from favor and the word "rhetoric" has acquired a faintly pejorative connotation.

Even if students were able to employ spoken language effectively, they would be hampered by lack of a descriptive language for recording or transmitting the information they would acquire. By way of contrast, when medical students are taught physical diagnosis, they are also taught a descriptive language that will serve to record their findings and communicate them to other doctors. An acutely arthritic joint is "red, hot, swollen, and tender." If I add "exquisitely tender," a physician might suspect infection in the joint, or gout. But if I write those findings—including "exquisite"—about the big toe, then a doctor would almost certainly think first of gout. That interpretation (often wrong) is implied by the word-picture. There lies a problem with any descriptive language: it should provide a representation, not an interpretation or conclusion. "Lemon-yellow" or "orangy-yellow" skin clearly picture different degrees of jaundice, and as such are much more descriptive than "light" or "deep" jaundice. I can go on to talk about wounds that are "red, swollen, and puckered around the sutures," pus that has a "fecal odor," the "sickeningly sweet smell of gangrene," or even "tympanitic abdomen" (one that produces a drumlike sound when tapped with a finger) to make the point that physicians' descriptive language for physical phenomena is rich and communicative, allowing the reader or listener to visualize what the observer is characterizing, without necessarily subscribing to the observer's conclusions. This richness of language is in sharp contrast to the poverty of words doctors use to portray persons. On their hospital charts, patients are often described for example, as "depressed"—these same patients never seem to be sad, gloomy, melancholy, unhappy, down-at-the-mouth, or even blue. The word "depressed" is not very descriptive; it is also a diagnostic term with a relatively precise meaning that often does not fit the patient so labeled.

Finding an appropriate language of description is not a minor matter. The genius of the disease theory is that it finally provided a

> . . . physicians' descriptive language for physical phenomena is rich and communicative . . . This richness of language is in sharp contrast to the poverty of words doctors use to portray persons.

basis for doctors to talk about sick persons, using a commonly agreed-upon language. Angina pectoris, coronary heart disease, rheumatic mitral valvular disease, oat cell carcinoma of the lung, and immune complex syndromes have common meanings wherever Western medicine is practiced because they are grounded in the anatomy and physiology of the body. The advantages that common terminology can provide for research and therapeutics cannot be overemphasized. It has taken one hundred and fifty years to achieve such unanimity of terms. (However, it requires frequent national and international conferences to maintain common language usage because of the natural drift in language practice.) Such linguistic precision became possible when doctors could agree not only on the words, but also on the defining characteristics of the diseases indicated by their names. Psychiatric nomenclature has suffered from lack of precision and agreed-upon definitions; these lacks are among the reasons for the deep schisms between different schools of psychiatry and between psychiatrists and non-psychiatrists. Finding a descriptive language for patients will certainly present difficulties, if we require total agreement on the definitions and nomenclatures portraying different types of people. It simply cannot be done.

I think the solution to the problem of describing persons for medical practice is to use everyday language. Unfortunately, this does not end the problem. To say, for example, that a patient is "nasty, churlish, and mean," or conversely, "a heaven-sent delight," certainly uses everyday language, but it does not solve the problem of definitions. One who is "churlish" to you may not be "churlish" in my eyes. I do not even believe we could all agree on the characteristics of "a heaven-sent delight." Such characterizations are often a matter of taste, or stem from life experiences. Because of the subjectivity of such words—for they are really words of opinion—they will not serve to describe persons for medical practice.

We need a model to point the way, and novelists would seem to be our best guides. Descriptive language is, of course, the writer's stock in trade. On the first page of Charles Dickens' *Dombey and Son*, Mr. Dombey and his newborn child are described:

> Dombey was about eight-and-forty years of age. Son was about forty-eight minutes. Dombey was rather bald, rather red, and though a handsome well-made man, too stern and pompous in appearance to be prepossessing. Son was very bald, and very red, and . . . somewhat crushed and spotty in his general effect. . . . On the brow of Dombey, Time and his brother Care had set some

marks. . . . Dombey . . . jingled and jingled the heavy gold watch-chain that extended from below his trim blue coat. . . .

In these few phrases we are given the outward appearance of Dombey, a London merchant. In addition we are told he is pompous. But Dickens does not simply say Dombey is pompous, he allows readers to come to the same conclusion by providing a word-portrait that fits a pompous man. Suppose, for instance, that Dombey had been "tall, fine-featured, with intense eyes that, like their owner, moved quickly. . . ." After such a depiction the author could not have said that Dombey was pompous; our own life experience would not support his conclusion. Thus an effective description, even when it contains an interpretation, provides the evidence to back it up. Here is a portrayal of Miss Tox:

> The lady thus presented was a long lean figure, wearing such a faded air that she seemed not to have been made in what linen-drapers call "fast colours" originally, and to have, little by little, washed out. But for this she might have been described as the very pink of general propitiation and politeness. From a long habit of listening admirably to everything that was said in her presence, and looking at speakers as if she were mentally engaged in taking off impressions of their images upon her soul, never to part with the same but with life, her head had quite settled on one side. Her hands had contracted a spasmodic habit of raising themselves of their own accord as an involuntary admiration. Her eyes were liable to a similar affection. She had the softest voice that ever was heard; and her nose, stupendously aquiline, had a little knob in the very centre or keystone of the bridge, whence it tended downwards towards her face, as an invincible determination never to turn up at anything.
> Miss Tox's dress, though perfectly genteel and good, had a certain character of angularity and scantiness. She was accustomed to wear odd weedy little flowers in her bonnets and caps. Strange grasses were sometimes perceived in her hair; and it was observed by the curious, of all her collars, frills, tuckers, wristbands, and other gossamer articles—indeed of everything she wore which had two ends to it intended to unite—the two ends were never on good terms, and wouldn't quite meet without a struggle. She had furry articles for winter wear, as tippets, boas, and muffs, which stood up on end in a rampant manner, and were not at all sleek. She was much given to the carrying about of small bags with snaps to them, that went off like little pistols when they were shut up; and when full-dressed, she wore round her neck the barrenest of lockets, representing a fishy old eye, with no approach

to speculation in it. These and other appearances of a similar nature had served to propagate the opinion that Miss Tox was a lady of what is called a limited independence, which she turned to the best account. Possibly her mincing gait encouraged the belief, and suggested that her clipping a step of ordinary compass into two or three originated in her habit of making the most of everything.

I do not know Miss Tox, but I think that she will not be the patient who openly argues with her doctor. It is possible, however, that in order to save money, she might not have a prescription filled unless her doctor had made its necessity absolutely clear (and even, perhaps, implied that not to get the medication and take it as directed would be an expression of disrespect). She seems a bit odd, however, and so one must be prepared for the unexpected and cautious in prejudging. But with the awareness created by what one *does* know, watching closely, asking careful questions, and listening attentively when she speaks, within a few visits the attentive physician will be able to take care of her very much better, and she will, with reason, feel understood.

Notice that much of what we have come to know of Miss Tox in these paragraphs comes from a description of her behavior—Miss Tox is pictured *in action.* If we were merely told that she is long and lean, or that her nose is "stupendously aquiline," such language, although unquestionably descriptive, would not reveal what kind of a person *she* is. Physical characteristics alone would provide an image not unlike early daguerreotypes—frozen and unnatural. The addition of behaviors rounds out the characterization. The conduct of Miss Tox that Dickens describes consists of bearing, demeanor, mannerisms, and habit of dress. From these small pieces of the total Miss Tox, we form an impression that supports speculation about some behaviors as quite possible for her and others as improbable. In general, even to know as little as we are told about Miss Tox provides an enormous amount of information about a total person. The reason is that people, in their dress, demeanor, gait, speech, facial expression, activities, work, in all of the characteristics that make up their persons, are *more*, rather than less, consistent. Let us see what the physician has the opportunity to observe, even in an initial interview: physical appearance, dress, cosmetics and ornamentation, demeanor, speech, gait, and manner of sitting. This is an impressive array of features—certainly enough to characterize the mode by which a person presents him or herself in a physician's office. (One must not forget that the presentation of self is, in part, context-dependent, and a doctor's office or a hospital is a very special context.) With the exception of physical appearance, each facet is a *behavior*, not an architectural detail—even facial expression is an action.

I am aware that great caution is necessary in interpreting and acting on the information provided by a patient's presentation and actions in the doctor's office. Some aspects of a person, such as sexual behavior, private fantasies, what is done in the intimacy of the home, or behavior during life-threatening illness, are not even open to educated speculation without much more knowledge of the individual, because it is in the nature of humankind that those behaviors may *not* be consistent with the manner in which the self is presented in everyday life. These aspects of the person *could* be known, if necessary, by asking about them. Further, interpreting behaviors can be open to many sources of error. Few would disagree, however, that accurate characterization of these aspects of human behavior would provide a good basis for representing an individual in words. The point is not whether offering an everyday language of description provides an infallible method of introducing the individual aspects of the sick person into the calculus of medical care, but whether teaching medical students and physicians how to describe the persons for whom they provide care will improve that care. What is required is that medical students be taught descriptive skills and learn a descriptive language. That is a task natural to the humanities.

Sometimes I hear the belief expressed that only science can truly measure and characterize. While it is true that measurement is a central task of science, there are problems of characterization that are, at least thus far, better solved by poets than scientists. The first stanzas of this poem by E.E. Cummings establish that a man is lying unconscious by the roadside. What is the matter with him? "swaddled with a frozen brook/ of pinkest vomit out of eyes/ which noticed nobody he looked/ as if he did not care to rise/ one hand did nothing on the vest/ its wideflung friend clenched weakly dirt/ while the mute trouserfly confessed/ a button solemnly inert." I cannot conceive of a more parsimonious or accurate characterization of the unconscious inebriate (Cummings, 1963). The point is not that science failed and poesy succeeded; it is that one cannot expect doctors to attend to sick persons as persons, if they cannot describe them. Neither science nor medicine has the descriptive tools, but the humanities do.

> **The point is. . . whether teaching medical students and physicians how to describe the persons for whom they provide care will improve that care. . . . That is a task natural to the humanities.**

Learning to Reason

The practice of medicine requires not only communication skills but also the ability to reason carefully and make decisions, capacities in which medical students receive little if any formal training. It is in this area of the liberal arts that philosophers have traditionally taught. Perhaps the best exemplar of what and how to teach medical students is K. Danner Clouser, a philosopher who has been teaching in a medical school for more than a decade. In a brief but rich essay, he describes clearly not only what philosophy has to offer in the medical setting, but how best to approach its teaching (Clouser, 1978).

Students are taught to formulate a clinical problem based on the information from the patient's history of the illness, physical examination, laboratory tests, X-ray examinations, and other studies. Once a problem has been formulated proceeding toward diagnostic or therapeutic action—based on both medical science and knowledge of the sick person rather than on a "recipe"—requires the ability to analyze the problem over a period of time and synthesize new information should the problem require reformulation. As a by-product of their education in the sciences, the analytic (reductive or scientific) thought that is employed in this process is familiar to most students by the time they have reached medical school. Active training is required, however, to improve their analytic skills and give them greater volitional control over their thinking. It is no small matter to be able to take a clinical problem apart, make appropriate distinctions, test for precision of thought, be able to explain the contribution of each aspect to the total problem, reconstruct the problem, and build in methods for further checking as the problem unfolds over time. Teaching such skills is both important and difficult. A clinical "case"—the story of a sick person—is like a text in many respects. It is precisely the ability to deal with temporal aspects of a problem unfolding over time (the sine qua non of a clinical problem) that marks good clinical thinking and that has been the downfall of computerized attempts at medical problem solving. (Other methods of analysis, such as text explication, that are familiar to humanists but virtually unknown in medicine, might be very useful devices in clinical practice.)

Medical students are generally more capable of thinking in a reductionist analytic manner then in a valuational mode. Valuational thought, in opposition to analytic thinking, is a synthetic, integrative, constructionist method, in which information is tested against previously held conceptions, meanings, and beliefs, and in the process is assigned meanings. For example, when a physician sees a patient with fever, cough, and chest pain, the diagnosis of pneumonia may come to mind. That process, of the conception "pneumonia" "coming to mind" to give cohesiveness to the diverse information, is valuational or synthetic thought. I believe that physicians use these two interdependent but competing modes of thought, analytic and valuational, without awareness and thus are unable to take full advantage of both. For contemporary medical students, analytic thought is more robust and well developed. Valuational thought, dealing more with the moral and the personal, is less developed, and more private. The inability to think in valuational terms as well as in analytic is unfortunate, for thinking of whole human beings in personal and moral terms requires valuational thinking. Human values cannot be arrived at by analytic thought. Thus, if students are to meet the needs of sick persons in terms of the personal and the moral (not only their diseases), they must also be able to think in valuational terms. Because it is less well understood, teaching valuational thinking is much more difficult than teaching analytic thought. One method is to provide "cases" replete with information about the person and the person's symptoms, and then ask the student to construct a story that tells in narrative form about the person *beyond* the information given. It is the effective use of narrative and metaphor that marks the ability to make "wholes" out of parts. Some stories will be truer to the information than others—increasing ability in valuational thought is marked by wholes that are truer to the parts. Such a method of teaching thinking allows what Robert Belknap has called the literary experiment to be brought to bear: what would happen to the story—and consequently to the patient—if some of the facts were varied? In the real world, facts are always missing, but in the literary experiment they can be provided, altered, or removed, allowing one to note the effects these changes would have in real life. It is valuational thought that most specifically allows the student's knowledge of everyday life and function to come into play in the care of patients. When both analytic (reductive) and valuational (synthetic) modes are brought together under conscious control,

> . . . physicians use these two interdependent but competing modes of thought, analytic and valuational, without awareness and thus are unable to take full advantage of both.

students are provided with powerful problem-solving tools.

One further goal in clinical thinking remains elusive. This is teaching physicians, or anyone else for that matter, to weigh equally in their thought processes value-laden or subjective, information and objective, or numerical, information. Until doctors can do this, until numerical data does not *always* win out over "softer," more subjective information, the goal of treating sick persons rather than diseases will remain unrealized. The greater confidence that observers place in numerical or other forms of "hard" data is one reason that such data overwhelm more subjective information. Too often, in coming to a medical decision, inappropriate numbers will be given greater credence than appropriate tactile, auditory, or visual information. In part, I suspect this bias is inevitable, and can only be overcome with experience. But one of the things that experience has to offer is confidence in the accuracy of sensory information. Such confidence could probably be speeded up by teaching students *how* to perceive, in much the same manner that graphic artists acquire such skills. Appropriately adapted to the clinician's tasks, the theory of artistic perception has much relevance to medicine (Berger, 1980).

While much of what I have discussed falls within the traditional purview of philosophy, some of the issues would require that philosophers in medical schools rethink their roles and develop new techniques for teaching. Here, as noted earlier with the other branches of humanism, the problem is to teach *medicine*, not philosophy.

Obstacles to a Role for the Humanities in Medicine Today

Obstruction to the entrance of the humanities into the medical curriculum takes two forms. The first is that which accompanies *all* changes in medical education, while the second seems specific to introducing the humanities.

First, the more general source of resistance. Medical schools have relatively fixed curricula. Students *must* take certain core curriculum courses which have precedence over other educational endeavors. Most laypersons will be pleased to know that students are required to study anatomy, physiology, biochemistry, microbiology, pathology, pharmacology, and similar disciplines. These "basic sciences" are, after all, the foundation of modern medicine. Since anatomists, biochemists, and pathologists are not unlike other educators, they believe that no student can ever learn enough of their subject. Consequently, most basic science departments are convinced that they do not have sufficient time in the core curriculum—they would do almost anything to get more teaching time. Almost no department *ever* voluntarily surrenders curriculum time—spouses and all earthly possessions might be relinquished, but curriculum time, never. It is obvious that if the humanities are to enter the core curriculum, something else has to go. Therefore, major curriculum reform is almost always necessary before humanities are successfully introduced into a medical school. This is one reason why the "new schools," established within the last two decades, have a greater commitment to the humanities than older medical colleges, for at the start there were no entrenched curricula that required changing.

Departments in medical schools are far more rigid and hierarchical than in most other colleges in a university. The chair of a department is a relatively permanent position that carries with it enormous authority over the faculty, space, and funds of that department. Beyond that, although relatively democratic governance procedures are in existence in most medical colleges, the faculty power structure is such that the clinical departments, especially medicine and surgery, have more power than the basic science departments. As with all stable institutions, however, the actual distribution of power can be quite different from the apparent distribution and will affect how change is carried out. Another source of resistance, albeit less open, is the fact that the introduction of the humanities into the medical curriculum represents the kind of change that may alter the balance of power. Another factor holding back the introduction of the humanities, or any other new material, is that the old established schools seem perpetually short of money and (especially) space—both of which are required when new subjects and faculty are introduced. Fortunately, humanists do not require laboratories, and this lessens their space requirements. I believe that this is one of the reasons why the increasing importance of the humanities in the southwestern medical schools has had little impact on the "establishment" schools. And of course, there is just plain inertia, friend of the status quo and enemy of change.

Much of the resistance noted above would disappear, I believe (given adequate funding and space), if humanists were content to teach an elective program and did not require core curriculum time. Funding agencies frequently do not believe that true change has taken place unless the core curriculum has changed. On the contrary, I think that there are *good reasons for not* teaching core courses, reasons that in some cases make electives more attractive and effective. In the first place, a medical school class numbers about one hundred students. Therefore, if they must all be taught—as is necessary in a core curriculum course—a lecture format must be chosen because of the few humanists on the faculty. Since lectures are not the best method of teaching most of the things that the humanities have to offer, a seminar format is often employed, which demands additional teachers. Such faculty frequently come from other departments, drawn by their commitment to the new program and their interest in the subject. When what is being taught is standard humanities fare, these informal faculty members (especially when drawn from the clinical departments) are often highly motivated and excellent.

If newer materials are to be taught, such as some of the subjects discussed in this monograph, then using outside faculty not specifically trained in and committed to the new methods runs the danger that the course will drift back toward the conventional. This "regression toward the mean" will be hastened if the students in the seminars are also not truly interested in the new approaches. Elective courses avoid both dangers. Enrollment can be kept small enough so that those who developed the course and its materials are the only teachers. The students who elect the courses will be the ones who are interested (the time has come to realize that medicine is not a unitary profession, and that what is appropriate training for clinicians may not be the best education for those heading for a career in basic research or in the specialities—radiology is an example—that do not involve direct patient care, and vice versa). If the course is good, the students will aid in its development and spread the word of its excellence. If the initial classes are a failure, total disaster is avoided and time is available for reconstruction. Gradually other faculty members will attend, and a cadre of teachers will develop who will help in the enlargement of the program. Over four or five years, a solid, well-tested, and successful program can be developed which may be *invited* into the core curriculum. In medical schools, *change takes time.* Of course a combination of approaches can be employed.

The resistances to change that I have described would be present no matter what curricular innovation threatened. There are, however, more important and more specific obstacles that impede the introduction of the humanities into medical school teaching. Medicine and the humanities—particularly philosophy—have had a longstanding love-hate relationship. In the document that is often taken as a symbol for the beginning of Western medicine, the Treatise on Ancient Medicine in the Hippocratic Corpus, the tension between the two is directly addressed and philosophy is attacked for being based on speculation. After a number of paragraphs, whose point is that, contrary to philosophy, the art of medicine should be based on *direct* observation of patients, their habits, diet and so on, the Hippocratic author says:

> Certain sophists and physicians say that it is not possible for anyone to know medicine who does not know what man is (and how he was made and how constructed), and that whoever would cure men properly, must learn this in the first place. But this saying rather appertains to philosophy, as Empedocles and certain others have described what man in his origin is, and how he was first made and constructed. But I think whatever such has been said or written by sophist or physician concerning nature has less connection with the art of medicine than with the art of painting. And I think that one cannot know anything certain respecting nature from any other quarter than from medicine. (Hippocrates, vol I, p. 143)

(For more on the history of the relationship between philosophy and medicine, see Pellegrino and Thomasma, 1981)

In the present context, it is useful to understand the belief that the antagonism between the humanities and medicine is a struggle between the "real" (medicine), and the "nonreal" (the humanities). Medicine deals with the realities of human biology, the body, and disease. It is (presently) founded on an atomistic rationalism, the reductionist methods of science, which have been enormously successful in first comprehending the mechanisms of disease and then providing tools for their relief. It views with suspicion and hostility attempts to understand these subjects based on pure hypothesis (from which sin it is not entirely free) and on methods that are not grounded first and last on objectively verifiable observation. The humanities, all of them, are seen by many physicians to suffer from both defects. Joanne Trautmann acknowledges the problem by defending the study of literature in medicine against the charge that medicine deals with the real world, but literature with a fictional one. She states: "In the ten years since I came to teach medical students, I have discerned this argument hidden in the statements of most of those who would keep literature out of medical education altogether or offer it only for a precious few" (Trautmann, 1982). Any person teaching the humanities in a medical environment will face the same bias. The objection cannot be dismissed; it must be acknowledged and dealt with.

If medicine dealt with the body, and the body *alone*, if what afflicted the sick person acted on the body and *only* on the body, and if it were possible to intervene in the illness without the interventions taking place in the person who has the illness, then believing that medicine, or doctors, deal only with the "real" might be possible (ignoring the difficulty with the word "real"). But as noted earlier, while medical science can abstract itself and deal solely with body parts, doctors who take care of patients do not have that luxury—they must work with people. Because of this, they are *always* faced with the *non*realities of their patients—the fears, desires, concerns, expectations, hopes, fantasies, and meanings that patients bring or attach to interactions with physicians—that *always* exist and that *always* influence their medical care. These "realities," often called "psychosocial issues," are, as noted earlier, better taught by literature and the other humanities.

The problem of dependence on speculation is not so easily dismissed. Nor is it much comfort that medicine itself, now and always in the past, has vehemently held positions that rest on beliefs about patients, diseases, and medicine that are not supported by observation. Issues such as the relation of mind and body, matters of causality, the place of spirituality and transcendence, the nature of the doctor-patient relationship, and ethics affect the practice of medicine, but are usually closed to resolution through direct observation. Despite the fact that the history of the profession's attempt to free itself from speculation goes back to Hippocrates, it must be emphasized, to medical students and faculty alike, that doctors cannot avoid such problems: they are inextricably bound up with the care of the sick. Instead, the choice is whether to employ the tools of the humanities that have been honed for centuries on just such questions, or to allow doctors to pretend the issues do not exist or use underdeveloped and sloppy thinking to address them. As Clouser has pointed out, although advancement in metaphysical questions is slow, because "these metaphysical questions are inevitable with every generation, it is important to have a skeleton crew of philosophers always on duty, if only to help each new generation to avoid the pitfalls, blind alleys, and philosophical howlers so well documented in the past" (Clouser, 1978).

Another conceptual problem underlies the gap between the

humanities and the sciences that, because it is more fundamental, may be of greater importance than the distinction between "real" and "unreal." This is the everyday viewpoint about the real world that has developed over these last three centuries of science and that is accepted as much by nonscientists as by scientists. The human body, as a collection of molecules, subcellular organelles, cells, and organs is conceived to be as subject to the mechanical laws of nature as are the rocks and the stars. These parts, the scientific conception continues, can be understood as separate objects virtually free of the interactive existence of each other and of the action of time. Thus, the objects of science, whether parts of the body or parts of the moon, are seen to be free of value and of quality. The paradox of the same real world is that when all the parts that make up an individual human come together, qualities and values are accepted as being the very essence of the resultant person, as is his or her capacity for self-determination. Before the parts come together they are the subject matter of science. When fully assembled and running under their own power, these persons are the subject matter of the humanities. Such a fundamental inconsistency in the thinking of a whole culture creates a rift between the humanities and the sciences that is not easily bridged. What reason is there to think that the problem can be solved in this era? In the simplest terms, medicine—a very pragmatic profession—*requires* the understandings, concepts, and the skills of the humanities if it is to progress, paradoxes notwithstanding. After all, if one paradox could be tolerated—and buried—in order that science could make the fantastic strides that it has accomplished, what is another small—and unspoken—contradiction. In more complex terms, the sciences, especially physics, because the world view inherited from a previous era is unsuited for further progress, create a pressure for new fundamental ideas that will surely lead to such change. However, the time scale of that kind of metaphysical change is so great that pragmatic professions move their actions well ahead of their professed beliefs and their language of discourse (Whitehead, 1967).

That progress occurs despite the inadequacy of fundamental ideas is shown by the fact that in the so-called "new" medical schools, those founded within the past fifteen or twenty years, the humanities have received much greater acceptance, as a necessary part of the medical student's education, than in the older, more conservative medical colleges. Greater acceptance is due, in part, to the fact that no entrenched curriculum priorities blocked their way. But more significant, I believe, is that these are the same schools that have stressed the importance of psychosocial issues, of treating the "whole person," of family practice, and of emphasizing health rather than merely the treatment of disease. Clearly, a commitment to these newer values in medicine will necessarily be accompanied by a commitment to the importance of the humanities in medical education.

Scholars of the humanities may not be thrilled at the idea of teaching the humanities as discussed here. Teaching in the medical environment does not promise the usual academic rewards, and the company and comfort of like-minded (and similarly trained) colleagues is absent. Since humanists' day-to-day work may diverge from their basic scholarly interests, publication becomes more problematic. Colleagues—peers and teachers alike—may look askance at their work, as though they were fallen gentry who had taken the easy way out by entering "trade."

Many in the humanities have followed the route of the sciences into analytic, reductionist methodologies and ways of thinking. To scholars who have embraced such ideas, the kind of humanities teaching that I have discussed here will seem old-fashioned, out of style, and uninteresting. In addition, I have repeatedly pointed out that to be successful in a medical school, teachers of the humanities must demonstrate to students the importance of their subjects *as medicine*. This is unattractive to faculty who were raised on teaching graduate students, or who believe that their subject matter is so intrinsically interesting and important that merely to make such knowledge available should guarantee student attention.

In the face of palpable distaste on the part of faculty in the humanities, it might seem unlikely that the humanities will ever enter contemporary medicine. To be crass, however, the desperate money problems of many university departments of humanities suggest that if there are funds for the humanities in medical schools, there will be humanities faculties in medical schools. The situation in philosophy is apposite. The impetus that initially propelled philosophers into the medical environment—into medical ethics and the philosophy of medicine—was a shortage of jobs in departments of philosophy and the availability of money and positions in medicine. But as pointed out earlier, interesting questions from medicine and the challenging immediacy of medicine's mission have enriched philosophy (Toulmin, 1982). I believe that the unresolved issues raised by the changing focus of medicine present the humanities with exciting challenges that can *only* be solved by the humanities. As with philosophy, the involvement of humanities scholars in medicine can serve to enlarge the humanities themselves. Teaching physicians how to deal conceptually with the issues discussed and providing skills and tools for action present research challenges to the humanities as well as to medicine.

> . . . the involvement of humanities scholars in medicine can serve to enlarge the humanities themselves.

New Challenges for the Humanities

Previously I presented the notion that the shift taking place in medicine is toward a focus on the sick person and away from an almost exclusive concern with the disease. More fundamentally, I believe, this change represents an increasing concern with wholeness and a turning away from atomistic thinking. However, it is one thing to be concerned with wholeness, but quite another to develop the intellectual tools to deal with it. As a cautionary tale, the holistic medicine movement provides almost a caricature of what happens when new concerns arise unaccompanied by new understandings or tools to deal with them. In its regard for the whole person, the movement has embraced unorthodox therapies (from acupuncture to zone therapy), more reductionist than orthodox medicine, and its wholehearted endorsement of things "natural" merely symbolizes a rejection of high technology but puts essentially nothing in its place.

There are corollaries of the interest in "wholes" that demonstrate some of the problems that must be solved before physicians can deal as effectively with whole persons as they do now with organs or diseased parts. I have chosen *ambiguity* as the first of these. It reveals not only how far is the route to be traveled from the current methods of medical science, but the contribution that must be made by the humanities. To make these points visually, I have used the device of showing color slides of a generous bouquet of flowers photographed daily from the time they were picked until they were utterly wilted, twelve days later. The group of flowers present a very complex image that changes in many different ways over time. By way of contrast, I show a similar set of slides of a group of three flowers photographed over the same time period. Although many things happen to the three flowers, the changes are easier to keep track of. In their reduced complexity, the three flowers are to the bunch, as understandings of body organs are to the whole person. Following the slides of the flowers are a set that show (only) the chest of a woman with cancer of the breast photographed daily from before her biopsy, to the bandaged torso following surgery, to the unveiling of the mutilated chest, to the evolution of the wound. Then pictures of her face over the same period are projected. The effect is striking. Medicine's view of cancer of the breast would appear to be captured by the pictures of the breasts. The deficiency of that perspective is driven home by the pictures of the woman's face. The face makes clear that science, as it is presently known in medicine, cannot cope with the ambiguity that appears to be the inevitable companion of wholeness. What is her face telling us? What does it mean? Based on what we see, what should we do? Is there only one answer to each of these questions? If there are more answers than one (and we know that to be true) are they the same for each observer, or does the whole of which we are speaking in medicine *necessarily* include the physician, or other caregiver as well as the sick person? Currently, to tolerate such ambiguity in science would be an error in reasoning. However, to eliminate the ambiguity in medicine eliminates the richness and subtle complexity that expands the meaning of every perception beyond the percept itself (Cassell, 1982). Alas, the previous sentence is, by now, almost a cliche, but it brings us no closer to teaching physicians how to deal with such ambiguity *as a basis for action*. However much difficulty doctors may have, literary and artistic critics deal with such ambiguity as their daily fare. Novelists (and critics) understand that if characters are presented for whom all meanings are clear and never in conflict, and who always speak with only one voice, such characters will be unreal and stick figure-like—caricatures of human complexity. Yet skilled writers are able to people their work with characters who are multivocal without being merely confusing, who live concurrently on more than one level without seeming strange, and in whom the triumph of the human spirit over the tragedy of sickness or death—such as one wants to see made possible by doctors' care of the sick—is seen to arise from the commonness of life's ambiguities rather than their rareness. So it *must* be the case that in those disciplines at least the groundwork has been laid from which a systematic understanding of ambiguity in medicine might emerge. Perhaps more than a groundwork exists, but what is lacking is the application to medicine and to the teaching of medical students. This is one of medicine's challenges to the humanities.

One does not need to go to the degree of complexity provided by the previous example to make clear medical deficiencies that are dealt with better in the humanities. For science, complexity is, in part, reduced by making sharp distinctions; putting forth definitions of terms and situations that allow of no overlap. A moment's

> **. . . to eliminate the ambiguity in medicine eliminates the richness and subtle complexity that expands the meaning of every perception beyond the percept itself.**

reflection, however, will show how artificial such distinctions often are. For example, when the flow of blood in vessels is smooth and orderly (the usual circumstance) it is called "laminar flow." Under certain conditions the flow becomes disorderly and that is called "turbulent flow." One can hear certain sounds over blood vessels in which the flow has become turbulent, and there may be other consequences as well. No matter how carefully "laminar flow" or "turbulent flow" are defined, however, it is inevitable that there is a point of "nondistinction," a state where one cannot make the distinction between the two kinds of flow—where it is not quite smooth, but not yet turbulent.

Medical scientists characteristically disregard such borderline states, as do the rest of us. They are put out of mind or defined out of existence because there are no good systems of thought to deal with them. Then we act as if nondistinction does not exist. The pretense that distinctions can always be made has been extraordinarily productive for medicine, but there are situations where such blinders become inadequate and we are brought up sharply against our *inability* to deal with nondistinction. Such a circumstance was brought clearly into focus when ethicists tried to define the characteristics of personhood that would provide a basis for deciding, say, when to turn off a respirator. If certain criteria of personhood were met, for example, "awareness of self" or "cerebral function," then the patient should be kept alive; if they were absent, then the patient might be allowed to die (Fletcher, 1972, 1974). It is obvious that no definitions of such terms can be made that will eliminate areas of nondistinction. Since doctors must act, and no action seems possible based on *non*distinction, the attempt to find precise "indicators of humanhood" fell by the wayside. But the beauty of medicine as a force for intellectual change is that the deficiency of the original solution was quickly apparent when doctors tried basing their actions on it. Further, even though the proposal did not work, the problem did not go away—decisions are still required about individual patients on respirators.

Phrases like "awareness of self" and "cerebral function" turn out to be metaphors (Cassell, 1976). Metaphors may be a problem for medicine, but they are no strangers in literature. Indeed, they have come under increasing study in recent years, and the fruits of that research may have applicability to medicine (Lakoff and Johnson, 1980). Thus, again, the move toward dealing with wholeness raises problems not at all foreign to the humanities, and in which progress in the humanities might be encouraged by the urgencies of medicine. I should point out that a move toward understanding nondistinction does not undercut the utility of clean distinctions. Newton's laws—quite satisfactory for everyday life—did not have to be abandoned when quantum physics solved problems for which Newtonian physics was inadequate. Similarly, one does not expect that solutions to issues of nondistinction that may arise from the humanities will displace from medicine the utility of scientific distinctions. Rather I believe these newer viewpoints will expand medicine's perspective and enhance physicians' ability to act.

Medical science and other sciences deal most often with static spatial concepts—with permanence and uniformity. On the other hand, to deal with whole persons, or any other living entity, necessitates coming to terms with the effects of the past, change, becoming, the unfolding novelty of the future, uncertainty, of time itself. Understanding illness and health as processes means developing new ways of seeing and understanding what is happening to patients. Because science *has* been so successful in structural terms, one can understand the reluctance to stipulate the universe of medicine in anything so seemingly shaky and evanescent as the language of process. But the humanities, from poetry to history, have had to find a language for change and have dealt with it sufficiently to provide at least a beginning for medicine's excursion into these new areas.

Understanding illness and health as processes means developing new ways of seeing and understanding what is happening to patients.

Conclusion

> ... in the decades ahead medicine will pursue scholars in the humanities until they and the humanities produce what medicine needs.

This essay has traced the long association of the humanities with medicine. It opened with the historical belief that the role of the humanities is to provide broadly educated physicians who, because of their background, can be expected to be more humane physicians. It closed by presenting a view of the future of medicine in which the participation of the humanities is necessary for progress to occur. Between the two roles is an entire spectrum of functions that can be provided only by the humanities. Many scholars may believe that *not one* of these roles is intrinsically interesting—that nothing challenging is offered to them personally, or to the humanities in general that would warrant leaving the secure, even insular, setting of university departments. On the other hand, universities in general and the humanities in particular are enough battered these days so that some teachers of the humanities may want to try out the medical environment if opportunities arise. I believe that those who do will find challenge aplenty.

However, whether the humanities are ready or not, we in medicine need new tools and skills, new insights and understandings to pursue our goals. Science cannot solve many of the problems that loom on medicine's horizon. Thus, like it or not, interested or not, in the decades ahead medicine will pursue scholars in the humanities until they and the humanities produce what medicine needs. Because it can be found nowhere else.

Bibliography

Berger, John. *About Looking*. New York: Pantheon Books, 1980.

Bertman, Sandra L. "Communication with the Dead: An Ongoing Experience as Expressed in Art, Literature, and Song." In *Between Life and Death*. Edited by Robert Kastenbaum. New York: Springer, 1979, p. 124 ff.

———. "The Language of Grief: Social Science Theories and Literary Practice." *Mosaic* 15/1:153–63, 1982 (Winter).

Bruer, John T. "Premedical Education and the Humanities: A Survey." Paper prepared for a Conference on the Humanities and Premedical Education, sponsored by The Rockefeller Foundation, February 25, 1980.

Burns, Chester. "Liberating Medical Minds: Can Historians Help?" In *The Role of the Humanities in Medical Education*. Edited by Donnie J. Self. Bio-medical Ethics Program, Eastern Virginia Medical School, Norfolk, Virginia, 1978.

Cassell, Eric J. "Moral Thought in Clinical Practice: Applying the Abstract to the Usual." In *Science, Ethics and Medicine*. Edited by Daniel Callahan and H. Tristram Engelhardt. Hastings-on-Hudson, N.Y.: The Hastings Center, 1976.

———. "The Nature of Suffering and the Goals of Medicine." *New England Journal of Medicine* 307:758–59, 1982.

Clouser, K. Danner. "Philosophy and Medical Education." In *The Role of the Humanities in Medical Education*. Edited by Donnie J. Self. Bio-medical Ethics Program, Eastern Virginia Medical School, Norfolk, Virginia, 1978.

Cody, John. "The Arts versus Angus Duer, M.D." In *The Role of the Humanities in Medical Education*. Edited by Donnie J. Self. Bio-medical Ethics Program, Eastern Virginia Medical School, Norfolk, Virginia, 1978.

Cummings, E.E. *Complete Poems 1913–1962*. New York: Harcourt Brace Jovanovich, 1963, p. 258.

Doi, Takeo. *The Anatomy of Dependence*. Tokyo: Kodansha International, 1981, p. 148.

Engel, George L. "Biomedicine's Failure to Achieve Flexnerian Standards of Education." *Journal of Medical Education* 53:387–92, 1978.

Fletcher, Joseph. "Indicators of Humanhood: A Tentative Profile of Man." *Hastings Center Report* 2:1–4, 1972.

———. "Four Indicators of Humanhood—The Inquiry Matures." *Hastings Center Report* 4:4–7, 1974.

Flexner, Abraham. *Medical Education in the United States and Canada*. A Report to the Carnegie Foundation for the Advancement of Teaching, 1910.

Hippocrates. *The Genuine Works of Hippocrates*. Translated by Francis Adams. 2 vols. New York: William Wood and Co, 1886.

Hudson, Robert. "Goals in the Teaching of Medical History." IV. *Clio Medico* 10:153–60, 1975.

Hunter, Kathryn. "Literature and Medicine: Standards for 'Applied' Literature." In *Applying the Humanities*. Edited by Daniel Callahan, Arthur L. Caplan, and Bruce Jennings. New York: Plenum Press. In press.

———. "Morehouse Human Values in Medicine Program 1978–1980: Reinforcing a Commitment to Primary Care." *Journal of Medical Education* 57:121–23, 1982.

Lain Entralgo, Pedro. "A Review of *Konzepte der Medizin in Vergangenheit and Gegenwart* by Karl Rothschuh." *Theoretical Medicine* 4:114–15, 1983.

Lakoff, George and Mark Johnson. *Metaphors We Live By*. Chicago: University of Chicago Press, 1980.

Larson, Magali Sarfatti. *The Rise of Professionalism: A Sociological Analysis*. Berkeley: University of California Press, 1977, p. 89.

McElhinney, Thomas K. *Human Values Teaching Programs for Health Professionals*. Ardmore, Pa.: Whitmore Publishing Co., 1981.

Marks, Sandy and Sandra Bertman. "Experiences of Learning About Death and Dying in the Undergraduate Anatomy Curriculum." *Journal of Medical Education* 55:48–52, 1980.

May, William. "A Public Justification for the Liberal Arts." *Liberal Education* 68:285–96, 1982.

Miller, Genevieve. "The Teaching of Medical History in the United States and Canada." *Bulletin of the History of Medicine* 43:344–75, 444–72, 553–86, 1969.

Mitchell, S. Weir. *Doctor and Patient*. 5th ed. Philadelphia: J.B. Lippincott, 1909, p. 56.

Moore, Anthony R. *The Missing Medical Text*. Melbourne University Press, 1978.

Olan, Levi A. "A Preliminary Summing Up." In *A Rational Faith*. Edited by Jack Bemporad. New York: Ktav Publishing House, 1977, p. 194.

Osler, William. *The Evolution of Modern Medicine*. New Haven: Yale University Press, 1919, p. xiv.

Pellegrino, Edmund. "The Clinical Arts and the Arts of the Word." *Pharos*, Fall 1981, pp. 2–8.

_____. "Educating the Humanist Physician." *JAMA* 227:1288–94, 1974.

_____. *Humanism and the Physician*. Knoxville, Tenn.: University of Tennessee Press, 1979.

_____. "The Humanities in Medical Education." *Mobius* 2:133–41, 1982.

_____. and David Thomasma. *A Philosophical Basis of Medical Practice*. New York: Oxford University Press, 1981. Chap. 1.

Percy, Walker. *The Message in the Bottle*. New York: Farrar, Straus and Giroux, 1975.

Peschel, Enid Rhodes, ed. *Medicine and Literature*. New York: Neale Watson Academic Publications, 1980.

Risse, Guenter. "Teaching Medical History in the 1970's: New Challenges and Approaches. II. *Clio Medico* 10:133–42, 1975.

Scarry, Elaine. "Work and the Body in Hardy and Other Nineteenth-Century Novelists." *Representations*. Summer 1983, no. 3, pp. 90–123.

Shortt, Samuel. "History in the Medical Curriculum." *JAMA* 248:79–81, 1982.

Thomas, Lewis. "Notes of a Biology-Watcher: How to Fix the Premedical Curriculum." *New England Journal of Medicine*, 298:1180–81, 1978.

Toulmin, Stephen. "How Medicine Saved the Life of Philosophy." *Perspectives in Biology and Medicine*, 24:736–50, 1982.

Trautman, Joanne. "The Wonders of Literature in Medical Education." *Mobius* 2:23–31 1982. p. 24.

_____. and Carol Pollard. *Literature and Medicine*. Philadelphia: Society for Health and Human Values, 1975.

Whitehead, Alfred N. *Science and the Modern World*. New York: The Free Press, 1967, p. 76ff.